YOUR KNOWLEDGE HAS VALUE

- We will publish your bachelor's and
  master's thesis, essays and papers

- Your own eBook and book -
  sold worldwide in all relevant shops

- Earn money with each sale

Upload your text at www.GRIN.com
and publish for free

Bibliographic information published by the German National Library:

The German National Library lists this publication in the National Bibliography; detailed bibliographic data are available on the Internet at http://dnb.dnb.de .

Imprint:

Copyright © 2016 GRIN Verlag, Open Publishing GmbH
Print and binding: Books on Demand GmbH, Norderstedt Germany
ISBN: 9783668330900

This book at GRIN:

http://www.grin.com/en/e-book/342696/chemistry-in-a-shopping-trolley-ascorbic-acid-concentrations-of-fresh

Kassidy-Rose McMahon

Chemistry in a Shopping Trolley. Ascorbic acid Concentrations of fresh orange, fresh lemon, store-bought orange juice, store-bought lemon juice and a Berocca tablet

GRIN Publishing

GRIN - Your knowledge has value

Since its foundation in 1998, GRIN has specialized in publishing academic texts by students, college teachers and other academics as e-book and printed book. The website www.grin.com is an ideal platform for presenting term papers, final papers, scientific essays, dissertations and specialist books.

Visit us on the internet:

http://www.grin.com/

http://www.facebook.com/grincom

http://www.twitter.com/grin_com

Contents

1.0 Abstract

The concentration of Ascorbic acid in lemons, oranges, lemon juice, orange juice and a Berocca tablet were calculated to determine the best way to receive the recommended daily intake of Ascorbic acid. This experiment was conducted to decide the best source of Ascorbic acid for pregnant women, which is available from a supermarket. The concentrations were finalised by titrating the five substances and then calculating the concentration of Ascorbic acid. It was found that lemons had the highest concentration of Ascorbic acid, which was 0.6175mol/L. The fresh lemon was followed by the Berocca tablet, which had a concentration of 0.585mol/L, lemon juice with 0.5625mol/L, orange with 0.105mol/L, and finally orange juice, which had 0.095mol/L. While the lemon had the highest concentration, it was decided that fresh oranges were the best source of Ascorbic acid for pregnant women. These findings are significant because many pregnant women do not receive the recommended daily intake of Ascorbic acid, which has detrimental health impacts on themselves, and their child.

2.0 Introduction

This task will undertake an experiment to determine and compare the concentration of Ascorbic acid in lemon, orange, lemon juice, orange juice and Berocca tablets to predict the best method for pregnant women to receive the recommended daily Ascorbic acid intake. Some commonly known sources of Ascorbic acid, more commonly known as Vitamin C, are natural orange juice, natural lemon juice, store-bought orange juice, store-bought lemon juice, and Berocca tablets. With the large variety of products claiming to contain the highly sought-after chemical Ascorbic acid, there are many misconceptions as to which products contain the highest concentration and the most cost-effective and beneficial to the human body. The experiment will be conducted to determine the best method for Ascorbic acid consumption during pregnancy.

3.0   Body

3.1   Acids and Bases

All substances are classified as either an acid or base, according to the three theories of acids and bases by Arrhenius, Brønsted-Lowry and Lewis. In the Arrhenius Theory of acids and bases, "Acids are substances which produce hydrogen ions in solution" and "bases are substances which product hydroxide ions in solution" (Clark, 2002). The process of neutralisation occurs when hydrogen ions and hydroxide are involved in a chemical reaction; in this reaction an acid and a base is mixed together. The product of this reaction is water and a salt. This chemical reaction is shown below:

Neutralisation according to Arrhenius' Theory

$$H^+_{(aq)} + OH^-_{(aq)} \longrightarrow H_2O_{(l)}$$

Example of Neutralisation Reaction

NaOH	+	HCl	→ NaCl	+	H_2O
Sodium Hydroxide	+	Hydrochloric acid	→ Salt	+	Water

(Study.com, 2016).

Strong and Weak Acids/Bases

The acidity and basicity of substances are determined by a pH Scale, which is numbered from 0 to 14, where substances ranked at 0 are extremely acidic, substances ranked at 14 are extremely basic, and substances ranked at 7 are neutral. Neutral substances are neither acidic, nor basic; examples of neutral substances are pure water and blood. The dissociation of ions in the solution determines the strength of the acid and base; for instance, an acid that has completely dissociated ions is referred to as a strong acid, meaning it is extremely acidic and appears towards the acidic end of the pH Scale. On the contrary, a basic substance that has only partially dissociated ions is referred to as a weak base, meaning it is not extremely basic and appears towards the neutral (middle) range of the pH Scale (Elmhurst College, 2003).

pH of Common Substances

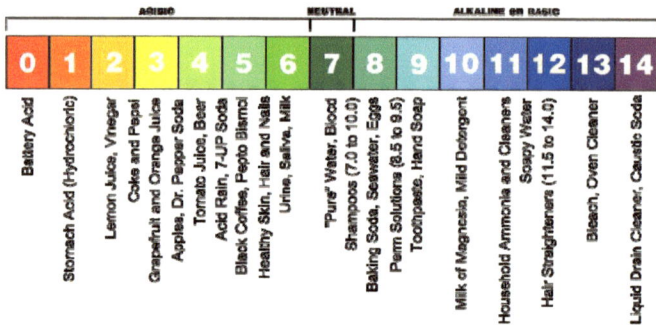

1: pH of Common Substances

Properties of Acids

According to the Arrhenius definition of acids and bases, there are different properties, which assist in determining whether substances are acidic or basic. The property already established in the Arrhenius Theory is the release of hydrogen ions into water (aqueous) solution, which classifies a substance as an acid. The Theory also states that acids neutralise bases in a neutralisation reaction; in this reaction the hydrogen ions of the acid and the hydroxide ions of the base combine to provide water and a salt as the product. Another distinctive property of acids is the result of testing the acidity of a substance with Litmus paper. When a sample of an acid is placed on Litmus paper, the paper transforms from blue to red, indicating that the substance is in fact an acid. The final property of acids is their sour taste; while many acids cannot be consumed due to their toxicity, humans can consume some weak acids, which provides a sour sensation to the taste buds (Chemtutor, 2013).

Arrhenius Acid Equation

HCl	+	H_2O	$\rightarrow$	H_3O+	+	Cl-
Hydrochloric acid		Water		Hydronium		Chlorine

Properties of Bases

Arrhenius also outlined the properties of bases in his Theory, which are still recognised and utilised today to identify basic substances. Arrhenius stated, "Bases release a hydroxide ion into water solution. Oppositely to the acidic properties outlined by Arrhenius, bases neutralise acids in neutralisation reactions, until a point of equilibrium is reached. Bases denature proteins, which explains the slippery textures of strong bases, such as cleaning products. As mentioned with the properties of acids, the basicity of substances can also be measured with litmus paper, but in the presence of a base, the litmus paper will change from red to blue, which is the opposite reaction of an acid. While very few food materials are basic and unfit for human consumption, bases have a distinctly bitter taste, unlike acids, which were stated to taste sour (Chemtutor, 2013).

Arrhenius Base Equation

NaOH	$\rightarrow$	Na^+	+	OH-
Sodium hydroxide		Sodium		Hydroxide

Brønsted-Lowry Theory of Acids and Bases

In the Brønsted-Lowry Theory of acids and bases, "An acid is a proton (hydrogen ion) donor" and "A base is a proton (hydrogen ion) acceptor" (Clarke, 2002). This theory involved conjugate acid-base pairs, which are "Two molecular species that easily transfer a hydrogen ion between them, especially from the acid to the base" (Dictionary.com). Brønsted-Lowry also discovered that substances could be amphoteric, meaning that they may act as an acid or a base in chemical reactions, as shown by the following reaction:

Amphoteric Substance

H_3O^+ ← Water accepts a proton, and is acting as a base. ← H_2O → Water loses a proton, and is acting as an acid. → OH^-

Brønsted-Lowry Equation
HCl donates a proton to water:

HCl	+	H_2O	$\rightarrow$	H_3O+	+	Cl-
Hydrochloric acid		Water		Hydronium		Chlorine

NH_3 accepts a proton from water:

NH_3	+	H_2O	$\rightarrow$	NH_4^+	+	OH-
Ammonia		Water		Ammonium		Hydroxide

(Chemed, 2016)

4

Lewis Theory of Acids and Bases

The third commonly recognised Theory of acids and bases is attributed to Lewis. Lewis stated that "An acid is an electron pair acceptor" and "A base is an electron pair donor" (Clarke, 2002).

Lewis acid-base Reaction

NaF	+	BF_3	$\rightarrow$	Na^+	+	BF_4^+
Sodium Fluoride		Boron Trifluoride		Sodium		Tetra fluoroborate

(Slideplayer, n.d.)

3.2   Le Chatelier's Principle

Neutralisation reactions occur when the concentration of an acid equalises the concentration of a base in an aqueous solution. Sometimes, a larger amount of acid or base is required in a neutralisation reaction if one of the substances has a higher or lower concentration than the other. The product of a neutralisation reaction is water and a salt, which indicates when the solution has reached chemical equilibrium. There is a chemical equilibrium "when the concentrations of reactants and products are in an unchanging ratio. Another way of saying this is that a system is in equilibrium when the forward and reverse reactions occur at equal rate" (Chemicool, 2014). This definition is supported is Le Chatelier's Principle, which was published in 1884 by French chemist and engineer Henry-Louis Le Chatelier.

Le Chatelier's Principle states, "A change in one of the variables that describe a system at equilibrium produces a shift in the position of the equilibrium that counteracts the effect of this change" (Chemed, 2016). The focus of Le Chatelier's Principle is the activities that a system undertakes when it temporarily shifts away from chemical equilibrium in order to return to its state of equilibrium. The three possibilities for a system to return to an equilibrium state, as explored by Le Chatelier were to alter the concentration of a component in the reaction, alter the pressure applied to a system, or to modify the temperature in which the reaction occurs. A system may be required to implement any of these strategies if the concentration of acid (or base) in a solution exceeds the concentration of base (or acid). For example, in Le Chatelier's Principle, when more H+ ions are produced by the acid, they will be neutralised by the same amount of base to reach equilibrium (Bhullar, 2016).

3.3   Titrations

"Titration is a general class of experiment where a known property of one solution is used to infer an unknown property of another solution" (Sparknotes, 2016). In a Titration setup, a Buret containing a base of known concentration is suspended above a flask or beaker containing an acid of unknown concentration. A stopcock is sealed at the bottom of the Buret to control the pressure of the base being released into the flask or beaker. The basic setup of a Titration is shown in the below diagram:

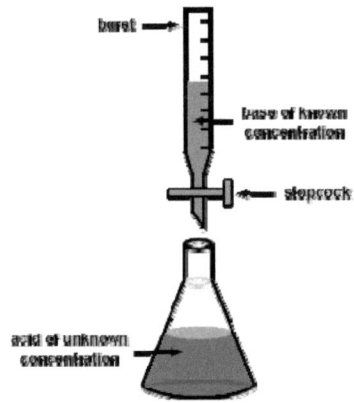

2: Titration Setup

There are two methods commonly used to determine when an acid has been titrated. The first method utilised a pH meter, which is placed in the acid on the unknown concentration. The base is gradually added to the acid until the pH reads exactly 7, which is known to be neutral on the pH scale. When this pH is achieved, the solution is no longer an acid, but it is not a base either; it is a salt in an aqueous solution. The second method involves an indicator, such as Phenolphthalein, which "is an acid or base whose conjugate acid or conjugate base has a colour different from that of the original compound" (Sparknotes, 2016). The first method is more effective because the exact point can be found exactly, but with the second method, the end point is likely to be missed.

3.4   Ascorbic acid

Ascorbic acid (Vitamin C) is a water-soluble vitamin, which is derived from foods, including citrus fruits, store-bought juices and vitamin supplements. While severe Ascorbic acid deficiency is rare, some symptoms include "dry and splitting hair; gingivitis (inflammation of the gums) and bleeding gums; rough, dry, scaly skin; decreased wound-healing rate, easy bruising; nosebleeds; and a decreased ability to ward off infection" (University of Maryland, 2013). Women require the most Ascorbic acid when breast-feeding, and when pregnant, which is the focus of this report. Women aged 18 years and younger require 80mg per day and women aged 19 and older require a minimum of 85mg per day. (Top 10 Home Remedies, 2016).

The molecular formula for Ascorbic acid is $C_6H_8O_6$ and its molecular weight is 176.12412g/mol. The chemical compound, as shown by the molecular formula, is comprised of Carbon, Hydrogen and Oxygen atoms. Its appearance can be described as "white to very pale yellow crystalline powder with a pleasant sharp acidic taste and almost odourless" (NTP, 1992). The pH of Ascorbic acid ranges from 2-3, meaning it is a fairly weak acidic, so it is fit for human consumption when in controlled concentrations. An overdose of Ascorbic acid may lead to diarrhoea, gastrointestinal irritation and predispose to renal oxalate calculi, though this condition is extremely rare (HSDB, 2016).

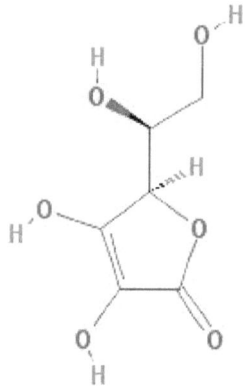

3: 2D Structure of Ascorbic acid

(PubChem, 2011).

3.5   Secondary Experiment

The aim of the secondary experiment by EasyChem was to "volumetrically analyse both freshly squeezed lemon juice and bottled lemon juice (through titration) and compare their citric acid contents" (EasyChem). In the experiment the Sodium Hydroxide base was standardised and the freshly squeezed lemon juice was titrated with Sodium Hydroxide, using Phenolphthalein indicator. The experiment was conducted three times with 1 rough trial and 2 accurate trials and the average of all trials were calculated.

Table 1: Raw Results of Secondary Experiment

Standardisation Titration			Freshly Squeezed Lemon Juice Titration			Bottled Lemon Juice Titration		
1 (Rough)	2 (Accurate)	3 (Accurate)	1 (Rough)	2 (Accurate)	3 (Accurate)	1 (Rough)	2 (Accurate)	3 (Accurate)
22.8 mL	22.7 mL	22.7 mL	25.5 mL	25.5 mL	25.5 mL	18.5 mL	18.3 mL	18.3 mL
Average (of 2 and 3)	22.7 mL		Average (of 2 and 3)	25.5 mL		Average (of 2 and 3)	18.3 mL	

(EasyChem).

The results showed that freshly squeezed lemon juice had the highest concentration Citric acid, followed by the standardisation titration and finally, the bottled lemon juice titration. This experiment was modified to test fresh lemon, fresh orange, lemon juice, orange juice and Berocca. It is expected that the Berocca will have the highest concentration, followed by the fresh lemon, fresh orange, lemon juice and orange juice. It is expected that the values will be near 25.5mL for the fresh fruits and 18.3mL for the store-bought juices, because of the results from the original experiment.

4.0   Experiment

4.1   Aim

The aim of the investigation was to determine and compare the concentration of Ascorbic acid in fresh lemon, fresh orange, orange juice, lemon juice and Berocca tablets to predict the best method for pregnant women to receive the recommended daily Ascorbic acid intake.

8

4.2   Hypothesis

It was hypothesised that the Berocca would have the highest concentration, followed by the fresh lemon, fresh orange, lemon juice and orange juice.

4.3   Justification of hypothesis

It was hypothesised that the Berocca tablet would contain the most Ascorbic acid because each tablet contains 500mg of Ascorbic acid, and the concentration of Ascorbic acid is equivalent to 7 oranges (Berocca Company). In the secondary experiment, the average amount of base required to titrate the fresh lemon was 25.5mL, so it is expected that the value will be similar. Lemons and oranges are both citrus fruits, so the results are expected to be quite close. The difference between the fresh and store-bought lemon juice in the secondary experiment was 7.2mL, so approximately the same difference is expected. It is believed that the store-bought lemon juice and store-bought orange juice will have approximately the same concentration.

4.4   Materials

1x	80mL Store-bought Lemon Juice
1x	80mL Store-bought Orange Juice
1x	1L 0.5M Sodium Hydrogen Sulphate (Standardised Solution)
1x	75mL Phenolphthalein Indicator
2x	50mL measuring cylinder
9x	Medium Oranges
9x	Medium Lemons
6x	Regular Berocca tablets
2x	Regular strainers
2x	Regular Buret stand
2x	Regular Buret

4x	Pair large gloves
4x	Pair regular safety goggles
2x	Regular Juice extractor
15x	250mL Flask
2x	Regular Funnel
1x	1L Distilled water
4x	150mL Beaker
8x	Regular Pipette
2x	Knife

4.5 Methodology

Step 1: The PPE was put on.

Step 2: The equipment was collected from the laboratory technician and cleaned thoroughly.

Step 3: The workspace was set up according to the *Experimental Set-up diagram.*

Step 4: The orange and lemons were cut in half with a knife.

Step 5: The juice from the oranges and lemons were extracted with a juice extractor.

Step 6: The juice extracted from the oranges and lemons were strained with the strainer.

Step 7: A Berocca tablet was placed in a 250mL flask with 20mL of distilled water. This step was repeated twice with two additional Berocca tablets.

Step 8: The Buret was filled with 50mL of Sodium Hydrogen Sulphate with a funnel.

Step 9: 20mL of natural orange juice was measured with a 50mL-measuring cylinder and poured into a 250mL flask.

Step 10: 3 drops of Phenolphthalein Indicator were added to the 250mL flask.

Step 11: The amount of Sodium Hydrogen Sulphate in the Buret was recorded.

Step 12: The flask was placed under the Buret.

Step 13: The nozzle on the Buret was turned so that the Sodium Hydrogen Sulphate gradually dripped into the flask.

Step 14: The flask was swirled constantly while the Sodium Hydrogen Sulphate poured into the flask.

Step 15: The pressure of the liquid was slowed as the colour change in the flask become more permanent.

Step 16: When the colour change in the flask was permanent, the nozzle on the Buret was turned off.

Step 17: The amount of Sodium Hydrogen Sulphate in the Buret was recording and special observations were noted.

Step 18: The liquid in the flask was poured down the sink.

Step 19: The flask and measuring cylinder was washed under the sink.

Step 20: The Buret was filled as necessary.

Step 21: Steps 8-19 were repeated with natural lemon juice.

Step 22: Steps 8-19 were repeated with store-bought orange juice.

Step 23: Steps 8-19 were repeated with store-bought lemon juice.

Step 24: Steps 8-19 were repeated with a Berocca tablet.

Step 25: Steps 8-19 were repeated twice with natural orange juice, natural lemon juice, store-bought orange juice, store-bought lemon juice, and Berocca tablets.

Step 26: All equipment was thoroughly cleaned and returned to the laboratory technician.

Step 27: The benches were cleaned with *Domestos Bench Spray*.

4.6   Experiment Set-up

4.7   Results

Fresh lemons

	Beginning	End	Difference
Trial 1	50mL	20mL	30mL
Trial 2	50mL	26mL	24mL
Trial 3	50mL	30mL	20mL
		Average	24.7mL

Fresh oranges

	Beginning	End	Difference
Trial 1	50mL	48mL	2mL
Trial 2	47.5mL	42.8mL	4.7mL
Trial 3	42.8mL	37mL	5.8mL
		Average	4.2mL

Lemon juice

	Beginning	End	Difference
Trial 1	50mL	25.5mL	24.5mL
Trial 2	50mL	30mL	20mL
Trial 3	50mL	37mL	23mL
		Average	22.5mL

Orange juice

	Beginning	End	Difference
Trial 1	28.8mL	25.5mL	3.3mL
Trial 2	21.7mL	17.5mL	4.2mL
Trial 3	21.7mL	17.5mL	4.2mL
		Average	3.8mL

Berocca

	Beginning	End	Difference
Trial 1	50mL	22.7mL	27.3mL
Trial 2	22.7mL	1mL	21.7mL
Trial 3	50mL	28.8mL	21.2mL
		Average	23.4mL

*The beginning and end are the burette values. The base used was NaOH.

5.0 Discussion

The aim of the investigation was to determine and compare the concentration of Ascorbic acid in fresh lemon, fresh orange, orange juice, lemon juice and Berocca tablets to predict the best method for pregnant women to receive the recommended daily Ascorbic acid intake. The objective of the report was achieved as the concentrations of each of the substances were found using the titration method and further calculations. It was hypothesised that the Berocca would have the highest concentration, followed by the fresh lemon, fresh orange, lemon juice and orange juice.

The product that had the highest Ascorbic acid concentration was the fresh lemon, which required an average of 24.7mL of base to be titrated and the calculated concentration of Ascorbic acid was 0.6175mol/L. In the secondary experiment, the fresh lemon required 25.5mL of base to be titrated, which is a different of 0.8mL. This result is quite close to the secondary experiment, so it is believed that the test was reliable. The difference in the volume could have occurred if different volume of fresh lemon was poured into the flask, or if the endpoint was misjudged in either experiment. The pH of a lemon is approximately 2 (The Chronicle Flask, 2013) meaning it is a rather strong acid. They also have a very sour taste, which is one of the key properties of acids. The presence of these properties means that it has a high concentration of acid (Ascorbic), which is why it is the most concentrated product that was tested in the experiment. The strengths of consuming lemons are that they have the highest concentration of Ascorbic acid, and therefore the amount needed will be lower. The limitations of consuming lemons are that they have an extremely sour taste and can cause heartburn in pregnant women (Livestrong, 2015). The hypothesis predicted that the fresh lemon would be the second most concentrated, but this part of the hypothesis was disproven.

The product that had the second highest Ascorbic acid concentration was the Berocca tablet, which required 23.4mL of base to be titrated and the calculated concentration of Ascorbic acid was 0.585mol/L. The Berocca tablet required 1.3ml less than the fresh lemon and the concentration was 0.0325mol/L lower. The lower concentration was believed to be because the pH of Berocca is 3.96 (Rushworth, 2010) and the tablet was dissolved in 20mL of distilled water. The pH of Berocca was 1.96pH less acidic, which could have caused the lower concentration. Some strengths of consuming Berocca is that only 1-2 tablets are required each day, the taste is not sour like fresh lemons and they are not know to cause heartburn in pregnant women. One of the limitations of Berocca is its side effects, including upset stomach, headache and unpleasant taste (Everyday Health, n.d.). The hypothesis predicted that the Berocca would be the most concentrated, but this part of the hypothesis was disproven.

The product that had the third highest Ascorbic acid concentration was the store-bought lemon juice, which required 22.5mL to be titrated and the calculated concentration of Ascorbic acid was 0.5625mol/L. The store-bought lemon juice required 0.9mL less than the Berocca tablet and 2.2mL less than fresh lemon. The store-bought lemon juice's concentration was 0.0225 lower than Berocca and 0.055mol/L lower than fresh lemon. The store-bought lemon juice contains additional ingredients, such as sugars, preservatives and additives, which enhance the product, but decrease the concentration of Ascorbic acid. The store-bought lemon juice had a lower concentration than the fresh lemon because of the added products, which reduce the costs of producing it. In the secondary experiment, the amount of base required to titrate the store-bought lemon juice was 18.3mL, which was 4.2mL (EasyChem). The gap in these results could be because different brands of lemon juice were tested, or the endpoint could have been missed because of very minute colour changes. Some of the strengths of store-bought lemon juice are tits enhanced taste (as opposed to fresh lemon) and the absence of side-effects (as found in Berocca). Some of the limitations, though, are that the concentration is lower than the fresh lemon and Berocca, and the added ingredients may not be the healthiest option for a pregnant lady (Oakenfull, 2014). The hypothesis predicted that store-bought lemon juice would be the fourth most concentrated, but this part of the hypothesis was disproven.

The product that had the fourth highest Ascorbic acid concentration was the fresh orange, which required only 4.2mL to be titrated and the calculated concentration of Ascorbic acid was 0.105mol/L. The fresh orange required 18.3mL less than lemon store-bought lemon juice, 19.2mL less than Berocca and 20.5mL less than fresh lemon. The fresh orange's concentration was 0.4575mol/L lower than lemon juice, 0.48mol/L lower than Berocca and 0.5125mol/L lower than fresh lemon. These results occurred because of the pH levels or fresh oranges, as opposed to fresh lemons and Berocca. The pH of fresh oranges ranges from 3.3 to 4.19 (Bruso, 2011), which is on average 3.745pH. The maximum pH of fresh orange is 0.23pH less acidic than Berocca and 2.19pH less acidic than fresh lemon. The strengths of fresh orange is that the taste is not overpowering and it is a natural fruit. A limitation is that it has the second lowest concentration of Ascorbic acid from the products tested. The hypothesis predicted that the fresh orange would be the third most concentrated, but this part of the hypothesis was disproven.

The product that had the lowest Ascorbic acid concentration was the store-bought orange juice, which required only 3.8mL to be titrated and the calculated concentration of Ascorbic acid was 0.095mol/L. The store-bought orange juice required 0.4mL less than fresh orange, 18.7mL less than lemon juice, 19.6mL less than Berocca and 20.9mL less than fresh lemon. The store-bought orange juice's Ascorbic acid concentration was 0.01mol/L lower than fresh orange, 0.4675mol/L lower than store-bought lemon juice, 0.49mol/L less than Berocca and 0.5225mol/L less than fresh lemon. The different between the fresh orange and store-bought orange juice is expected to be because of the added ingredients, as predicted with the fresh lemon and store-bought lemon juice. The strengths of store-bought orange juice are that there is no sour taste and there many different flavour combinations. The limitations of store-bought orange juice are that it has the lowest Ascorbic acid concentrations of all tested products, and the added ingredients could reduce the healthiness of the mother and child.

Some errors, which may have occurred during the experiment, are missing the end point of the reaction, using the wrong amount of indicator and decreased volume as a result of a leaking Buret (ChemBuddy, 2011). A common mistake made when performing titration is missing the endpoint of the titration. This occurs when the colour change has not been recognised or if the flask was not swirled thoroughly. The opposite may occur if the temporary colour change was mistake as permanent, meaning the endpoint had not yet arrived. The amount of indicator used in a titration can shift the endpoint, so if too much or too little indicator was dripped into the flask, the endpoint may have been delayed or premature. This error could occur if the cap of the indicator bottle was not tightened sufficiently or if the bottle was squeezed too hard. Another error, which could have occurred, is the recording of the incorrect volume as a result of a leaking Buret. After the Buret was filled, some of the liquid could have leaked from the bottom before the flask was placed underneath. If the Buret was left unattended for a small amount of time, the amount of liquid that leaked could have been quite significant.

6.0   Recommendations

To improve the experiment it is recommended that a digital pH measure be used to identify the endpoint of a reaction so that the results are completely accurate. Another suggestion is that caution be taken when applying the indicator to the acidic solution. Prior to an experiment, all equipment should be checked to be in working order to ensure that Burets do not leak. If liquid has leaked from the Buret before the flask has been placed underneath, more correct volume of the Buret should be recorded. These recommendations will improve the accuracy of the results and therefore the reliability of the data. This will make the data more useful and relevant when making recommendations to meet the aim of the investigation.

To expand on this experiment more substances could be tested, such as different fruits and vegetables, store-bought drinks and different brands of supplements. Different vitamins and minerals required by pregnant women could also be tested, such as Calcium, Vitamin D and Iron. Products aimed at different age groups, including infants, children, adolescents, adults and elderly people could also be tested to determine the best-suited products for different groups of people. The tests should be repeated more times to guarantee the accuracy and therefore the reliability of the data being presented. These are some opportunities to enhance the complexity of the experiment and further scientific knowledge.

It is recommended that women fresh oranges during their pregnancy as their source of Ascorbic acid. Even though fresh lemon had the highest concentration, heartburn is a likely side effect of constantly consuming the product. While Berocca had the next highest concentration, it had unfavourable side effects including upset stomach, headache and unpleasant taste. Store-bought lemon juice had the next highest concentration but the added ingredients, such as additives, preservatives and sugars are not ideal or pregnant women and their children. The same principle applies to store-bought orange juice, which means that fresh orange is the best source of Ascorbic acid for pregnant women.

7.0   Conclusion

This report has compared the performance of different products in delivering the necessary amount of Ascorbic acid to the bodies of pregnant women. The finding of this report was that fresh orange is the best source of Ascorbic acid for pregnant women. Other options, including Berocca, fresh lemons, store-bought orange juice, and store-bought lemon juice could also be suitable substitutes, as they too have adequate concentrations of Ascorbic acid. It is believed that the information provided in this report is accurate and reliable as the results gathered from the experiment are supported by primary and secondary data. Several recommendations have been provided to expand on the experiment so that more conclusions can be drawn about the effectiveness of supermarket products in providing the necessary vitamins and minerals to different groups of people so that they remain healthy.

8.0 Bibliography

10 Best Natural Sources of Vitamin C. (2016). Retrieved August 6, 2016, from Top 10 Home Remedies: http://www.top10homeremedies.com/kitchen-ingredients/10-best-natural-sources-of-vitamin-c.html

Acids and Bases. (1997-2013). Retrieved August 11, 2016, from ChemTutor: http://www.chemtutor.com/acid.htm

Agriculture, U. S. (2016, May). *Basic Report: 09150, Lemons, raw, without peel.* Retrieved August 13, 2016, from Agricultural Research Service: https://ndb.nal.usda.gov/ndb/foods/show/2255?fgcd=&manu=&lfacet=&format=&count=&max=35&offset=&sort=&qlookup=09150

Ascorbic acid. (n.d.). Retrieved August 5, 2016, from PubChem: https://pubchem.ncbi.nlm.nih.gov/compound/ascorbic_acid#section=Top

Clarke, J. (2013, November). *Theories of Acids and Bases.* Retrieved August 12, 2016, from Chem Guide: http://www.chemguide.co.uk/physical/acidbaseeqia/theories.html

Definition of Equilibrium. (2014). Retrieved August 9, 2016, from Chemicool: http://www.chemicool.com/definition/equilibrium.html

Health, N. I. (2011, June 24). *Vitamin C.* Retrieved July 30, 2016, from Office of Dietary Supplements: https://ods.od.nih.gov/factsheets/vitaminc-consumer/

How Much Vitamin C Is in an Orange? (2016, August 14). Retrieved August 14, 2016, from New Health Guide: http://www.newhealthguide.org/How-Much-Vitamin-C-Is-In-An-Orange.html

Le Chatelier's Principle. (n.d.). Retrieved August 8, 2016, from Chemed: http://chemed.chem.purdue.edu/genchem/topicreview/bp/ch16/lechat.html

Lesson 1.5: The pH Scale. (2016, March). Retrieved August 10, 2016, from Reddit: https://www.reddit.com/r/AsianBeauty/comments/44ivss/skin_basics_15_the_ph_scal e/

Maryland, U. o. (2016). *Vitamin C (Ascorbic acid).* Retrieved August 6, 2016, from University of Maryland Medical Centre: http://umm.edu/health/medical/altmed/supplement/vitamin-c-ascorbic-acid

Nave, R. (2012). Retrieved July 29, 2016, from Hyper Physics: http://hyperphysics.phy-astr.gsu.edu/hbase/chemical/acid2.html

Nave, R. (2012). Retrieved July 27, 2016, from Hyper Physics: http://hyperphysics.phy-astr.gsu.edu/hbase/chemical/acidbase.html

Products. (2015, August 12). Retrieved August 15, 2016, from Berocca: http://www.berocca.com.au/products/berocca-performance/

Titration Errors. (2011, May 25). Retrieved August 1, 2016, from Titrations.info: http://www.titrations.info/titration-errors

Titration Experiment. (n.d.). Retrieved August 3, 2016, from EasyChem: http://www.easychem.com.au/the-acidic-environment/acid-base-definitions/titration-experiment

Titrations. (2016). Retrieved August 7, 2016, from SparkNotes: http://www.sparknotes.com/chemistry/acidsbases/titrations/section1.rhtml

Vitamin C in your pregnancy diet. (2016, June). Retrieved August 4, 2016, from Baby Centre: http://www.babycenter.com/0_vitamin-c-in-your-pregnancy-diet_660.bc

Weil, A. (2012, October 29). *Vitamin C Benefits.* Retrieved July 28, 2016, from WEIL: http://www.drweil.com/drw/u/ART02811/facts-on-vitamin-c

Your Healthy Pregnancy Diet: Top Nutrients. (2005-2016). Retrieved August 2, 2016, from WebMD: http://www.webmd.com/baby/pregnancy-diet-nutrients-you-need#1

9.0 Appendix

Fresh lemon calculation

M1 = Concentration of acid   ?

V1 = Volume of acid   20mL

M2 = Concentration of base   0.5M

M2 = Volume of base   24.7

$$x(M1) = \frac{0.0247 * 0.5}{0.02}$$

$$M1 = 0.6175 mol/L$$

Fresh orange calculation

M1 = Concentration of acid   ?

V1 = Volume of acid   20mL

M2 = Concentration of base   0.5M

M2 = Volume of base   4.2

$$x(M1) = \frac{0.0042 * 0.5}{0.02}$$

$$M1 = 0.105 mol/L$$

Lemon juice calculation

M1 = Concentration of acid   ?

V1 = Volume of acid   20mL

M2 = Concentration of base   0.5M

M2 = Volume of base   22.5

$$x(M1) = \frac{0.0225 * 0.5}{0.02}$$

$$M1 = 0.5625 mol/L$$

Orange juice calculation

M1 = Concentration of acid   ?

V1 = Volume of acid   20mL

M2 = Concentration of base   0.5M

M2 = Volume of base   3.8

$$x(M1) = \frac{0.0038 * 0.5}{0.02}$$

$M1 = 0.095 mol/L$

4.8g Berocca Tablet dissolved in 20ml of distilled water

M1 = Concentration of acid   ?

V1 = Volume of acid   20mL

M2 = Concentration of base   0.5M

M2 = Volume of base   23.4mL

$$x(M1) = \frac{0.0234 * 0.5}{0.02}$$

$x(M1) = 0.585$